The Greatest Science Challenge
in the History of Civilization

by Rolf A. F. Witzsche

Contents

About the Illustrated Science series:
Ice Age – Climate Change
and the book
The Greatest Science Challenge in History

Numerous fields of evidence tell us that the next Ice Age is near. Most of the evidence was discovered in the 1990s and thereafter. Some evidence is measured in ice cores; some is measured in space, by satellites. Some measurements are also made on the ground in terms of measurements of the Earth's magnetic-pole drift observed in northern Canada. All of this is seen combined with high-energy physics experiments at a leading national laboratory, and is also explored in the small in static experiments.

Against the background of these widely diverse types of evidence that have been recently discovered, the historic Little Ice Age in the 1600s, takes on a new dimension as a yardstick for measuring the future that by this evidence promises to be up to 40-times colder than the Little Ice Age had been. The evidence poses a challenge, the greatest of all times.

In the Little Ice Age between 10% and up to 30% of the populations in Europe had perished by starvation. The last Big Ice Age was evidently vastly harsher. Only 1-10 million people emerged from it alive. That's all we had after 2 million years of development. We want to do far better this time around; and we can, with large-scale technological infrastructures for our food supply. But will we create them? Will we get the job done in the 30 years that we still have left before the Ice Age starts anew? And how certain are we that the phase shift to the next glaciation period will begin in the 2050s? We have 58 items of evidence to support this as a possibility. But can we move with the evidence?

It takes an independent researcher to brake the taboos that have kept mainstream cosmology imprisoned, increasingly, during the past century, even while what is regarded as taboo is known to be wrong.

High-resolution color images, of the images in this book, can be obtained at www.iceagetheatre.ca

The Greatest Science Challenge in the History of Civilization

The largest, impending, physical event

The challenge is, to respond in the most efficient manner to the largest, impending, physical event in the history of civilization, in the form of the start of the next Ice Age with the Sun going inactive sometime in the 2050s timeframe.

The challenge is so great

The challenge is so great that it will determine the future existence of all the northern nations on Earth, like Canada, Russia, most nations of Europe, some parts of Asia, and to a large measure the USA.

Living in a world with an inactive Sun

The challenge of living in a world with an inactive Sun for long periods, with 70% less energy being radiated by it, is so enormous and far reaching in is consequences that it seems hard to accept as our potential future in the near term, even while volumes of evidence stand in support of it.

Milankovitch Cycles theory discredited

It has long been recognized that enormous ice ages have occurred in the past, with ice sheets more than 10,000 feet deep accumulating across the northern contents, but for the lack of any real plausible theory for the cause for such events, with the Milankovitch Cycles theory discredited, the subject of the ice ages has been largely pushed into the background as an academic concern that is pertaining to distant future times, instead of to the immediate present with enormous global consequences.

Institutions of science have a critical role

The world's institutions of science have a critical role to play, therefore, in the resulting arena of critical superlatives. By definition science is focused on discovering what is real, especially the aspects that are invisible to the senses, but are discernable in the mind by their effects.

Science redirected

CERN - CLOUD project - Jasper Kirkby

Unfortunately, the world of science is not that simple. In exploring the unknown, numerous potential avenues are typically pursued, with hypothesis that often fall by the wayside as invalid. In addition, the orientation of science is all too often also redirected in the service of political objectives, since politics provide the funding, which makes the discovery of truth even more difficult, especially at the leading edge where conflicting interests and numerous conflicting ideologies, converge.

A momentous Science Challenge

David LaPoint - The Primer Fields

The Ice Age Challenge is presently emerging at the center of the self-conflicting landscape of opinions, doctrines, and guided objectives. It has been put on the map once again as the result of some critical scientific discoveries, which, by their nature render the dawning Ice Age Challenge, a momentous Science Challenge. The challenge is momentous, because of the enormous consequences involved.

Ice Age Challenge is a Science Challenge

The Ice Age Challenge is a Science Challenge of the highest order, because the potential consequences for humanity and civilization, if the challenge is not met, pale the worst that has been experienced in known history, into insignificance.

Leading edge discoveries of universal physical principles

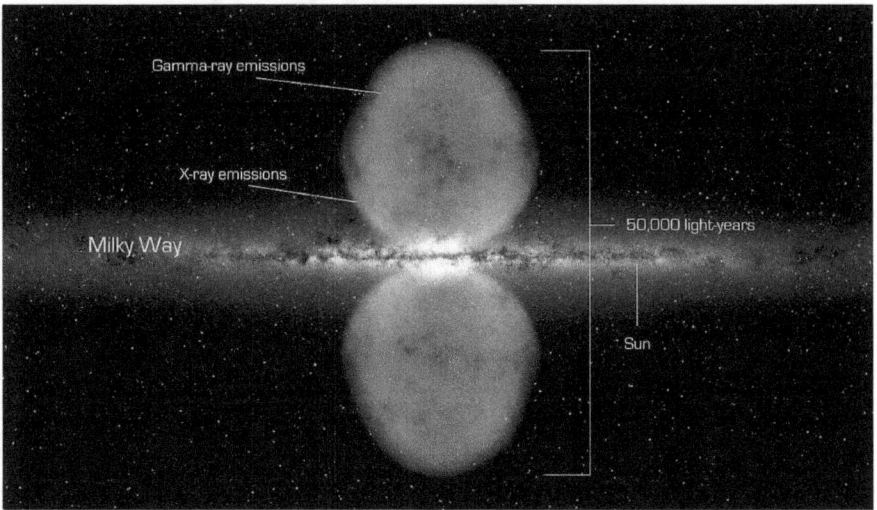

The great challenge has become immediate in recent years by a number of leading edge discoveries of universal physical principles, which once they are more fully understood, provide a new basis for solutions to a wide range of 'difficult' problems that have 'mudded the water' in a number of scientific disciplines.

Conflicts at the leading edge of science

Many of the conflicts at the leading edge of science, especially in areas pertaining to the Ice Age Challenge, have their roots in a single controversy. The controversy is, whether cosmic space is pervaded by streams of electrically charged particles, called plasma, that link and power all galaxies in the universe and their stars, and determine their interactions, or whether the universe is a sea of isolated, self-powered entities, interacting loosely with no greater force than the force of gravity.

Science controversy has deep roots

ESO/VIMOS galaxy cluster ACO 3341

This science controversy has deep roots and remains strong, even while evermore evidence is becoming recognized in support of the plasma-universe concept. The problem with interstellar plasma is that its presence is invisible, because of the small size of its particles that are 100,000 times smaller than the smallest atom. Plasma can only be seen by its effects on the atoms in space when the plasma current is strong, and by the dynamic effects of the electric force that becomes visible by the often string-like alignment of galaxies and planets.

While the controversy over the existence of plasma in space appears to be purely academic in nature, in real terms the opposite is true, because the future existence of humanity depends potentially on how this controversy is resolved. This is so, because the recognition of the nature of our Sun is tied up in this controversy, in which the Ice Age Challenge is located, that thereby has become the greatest challenge in science. The future existence of us all, in the 'near' term, will be determined by how the great Science Challenge will be resolved and be responded to, and by whom, and if indeed it will be responded to at all.

The 1970 Nobel Prize in Physics

The Sun may be seen in part as a MHD system

Hannes Olof Gösta Alfvén
electrical engineer,
plasma physicist
1970 Nobel Prize in Physics
for his pioneering work on
magnetohydrodynamics (MHD)

SDO/AIA 171 2013-04-09 14:15:12 UT

As I hinted at earlier, some parts of the Science Challenge are not new. One of the deep historic threads leads to the Swedish electrical engineer, plasma physicist, and winner of the 1970 Nobel Prize in Physics, Hannes Alfven, who became famous for his work in plasma physics and for his work in the related subject of magneto hydrodynamics.

Exploring the dynamics of plasma in space

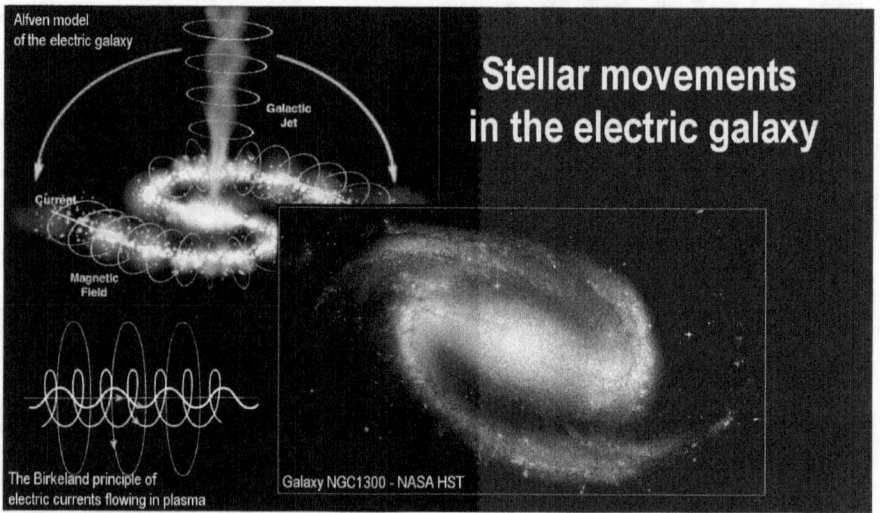

Hannes Alfven also became famous for his work in exploring the dynamics of plasma in space as an organizing force for the functioning of galaxies.

Alfven eclipsed by the Big Bang Creation theory

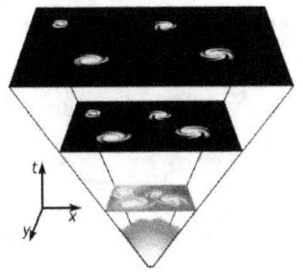

The Big Bang creation myth refuted by the electric solar fusion model

Alfven's breakthrough work was quickly eclipsed in his time, and pushed into the background, by the Big Bang Creation theory, of the universe, which had been strongly promoted at the time as a counter-ideology that is based on the assumption that plasma in space does not exist as a causative or organizing force. However, when one puts the omitted plasma back into the concepts regarding the nature of the universe, the fabric of the Big Bang Creation theory becomes suddenly very thin and full of holes, in the light of the resulting evidence.

Opposite of the Big Bang Creation theory

Researchers at Los Alamos National Laboratory in the USA, go into the opposite direction of the Big Bang Creation theory.

Electric plasma streams as the main organizing force

The Capodimonte Deep Field - of 35,000 galaxies

NASA

Researchers at the laboratory came to the recognition that 99.999% of the mass of the universe exists in plasma form, primarily in the form of electrons and protons that carry an electric 'charge' and pervade all space, free flowing, which by their electric potential become self-organized into vast electric plasma streams as the main organizing force of the universe in intergalactic space, some extending across billions of light years.

Experiments in high-energy physics

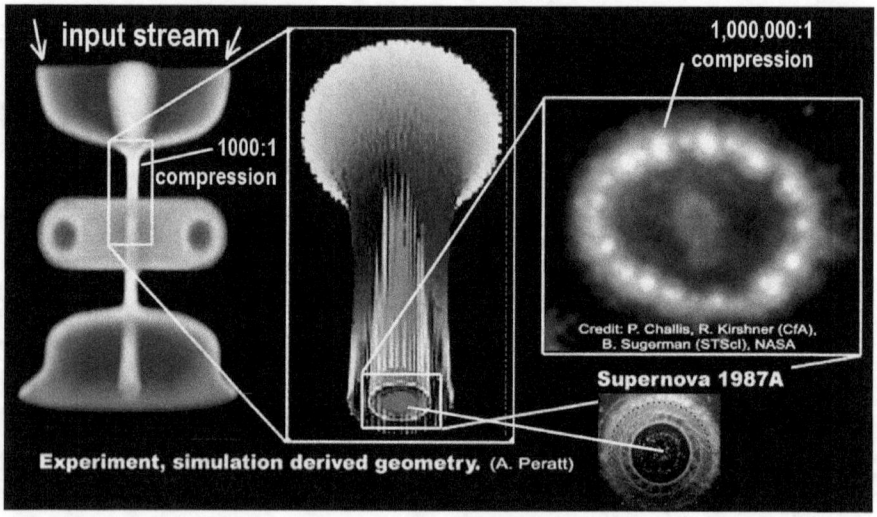

Extensive experiments in high-energy physics provide visible evidence in the form of unique shapes of geometry, which indicate that controlling cosmic forces of similar geometry are electric in nature for which numerous forms of similar geometry are evident, in space, though much larger in scale.

The researcher Anthony Peratt

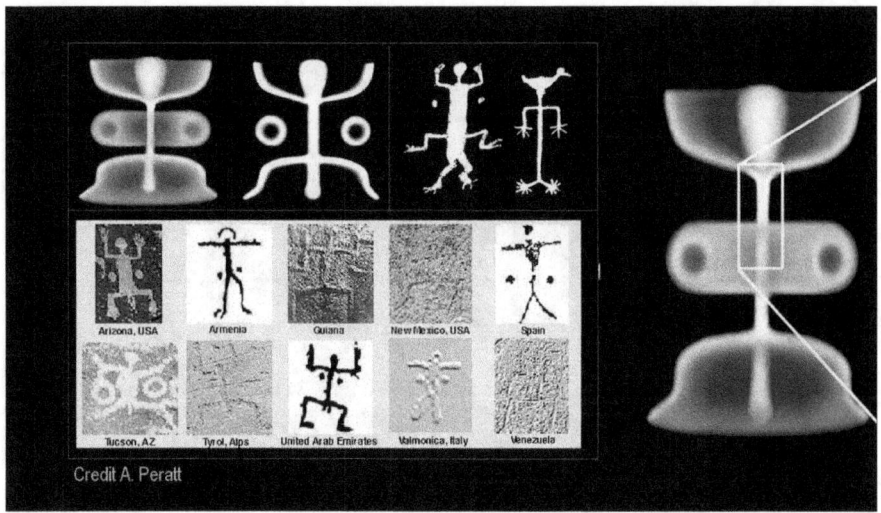

Credit A. Peratt

The researcher Anthony Peratt also discovered evidence in archetypal drawings from widely dispersed regions of the world, that basically reflect the same plasma-flow geometry. He suggests that large plasma-flow structures would likely have once been visible in ancient skies, probably during the electrically 'hot' times of the interglacial maximum period.

Thunderbolts Project

The Thunderbolts Project

NASA immage

Large-scale effects of cosmic electric actions between stars and planets have been discovered by researchers all over the world, with a flood of evidence being presented that we live in an electric universe. Many of the research findings are published in the extensive archive of the "Thunderbolts Project."

Electric forming of the Grand Canyon

Most of the project's research papers are focused on large-scale historic, electric astrophysical events, such as the forming of the Grand Canyon in Arizona, and of the gigantic canyon of Valles Marineris on Mars. The Sun, however, is rarely included as an electric astrophysical phenomenon. However, it is here where the critical consequences for our time, of electro astrophysics is located.

The 'weight' of the science exploration

© Milloslav Druckmueller/Barcroft

http://www.zam.fme.vutbr.cz/~druck/Eclipse/ - an example of the amazing solar eclipse photography
of Milloslav Druckmueller

The inclusion of the Sun into the sphere of the electric universe, takes the 'weight' of the science exploration out of the purely academic realm, into the practical realm of down-to-Earth living and frontline politics, where the consequences weigh in enormously, if the Science Challenge is not met and the potential consequences are not prepared for.

Researchers at the Los Alamos National Laboratory

Researchers at the Los Alamos National Laboratory, too, keep the Sun out of sight as an electric plasma object, possibly for political reasons. Nevertheless, the work accomplished there, at the labs, has opened the door to further breakthroughs by other researchers.

Plasma researcher David LaPoint

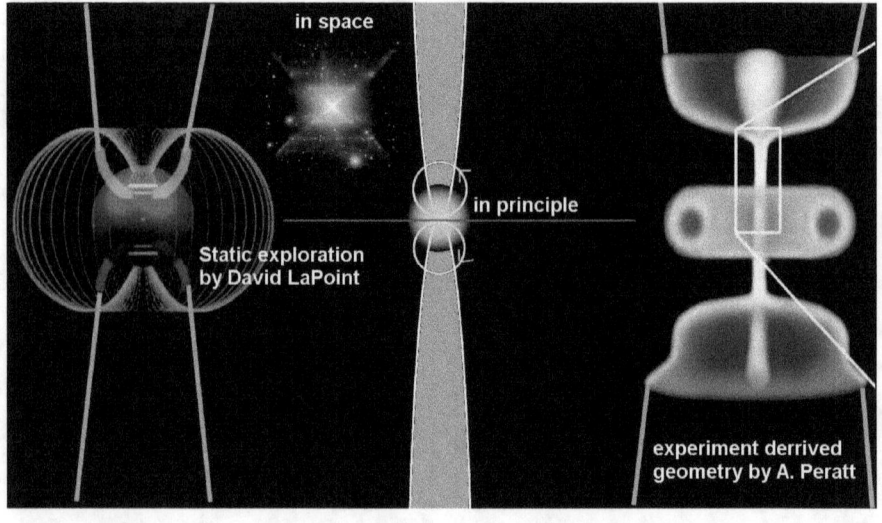

The plasma researcher David LaPoint took the critical step of putting the Sun back into the electric universe, as an electric component. He brought the Electric-Sun concept back into few, which from the time of the eclipse of Alfven, had been kept largely out of the sphere of serious science explorations.

David LaPoint utilized partially static experiments to explore the operating principles of the unique magnetic geometry observed in high-power dynamic experiments. He termed the operating dynamic fields, the "Primer Fields." His exploration work is important, as it provides a firm foundation for the long standing theory of the plasma Sun, which puts the theory back onto the map, but with a twist. The old theory now became relevant again, with reflections of the discovered principles becoming visible throughout the universe.

Since plasma flows can be seen in space by their effect on atomic gases, in areas of high-density plasma streams, the operating principles that have been explored in laboratory experiments, are suddenly being recognized in space, often clearly visible, even while plasma itself is too minuscule in size to be directly observable.

Minimal plasma density for the Primer Fields

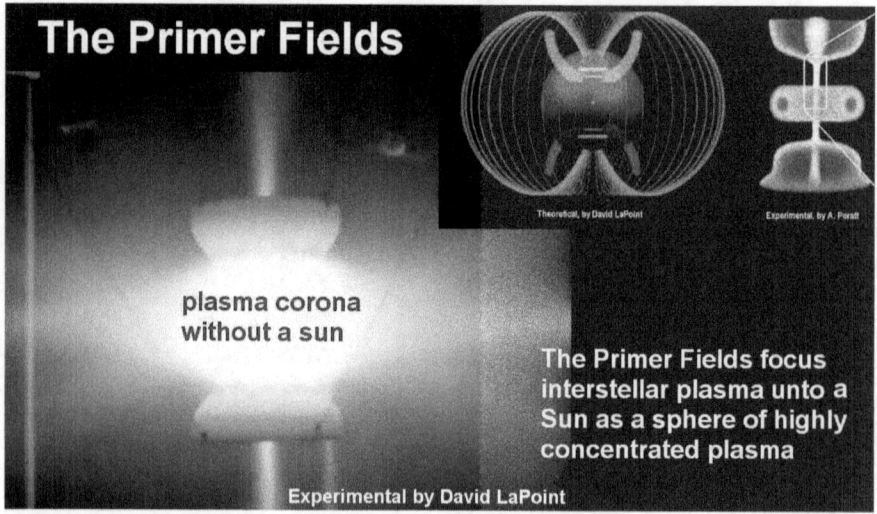

LaPoint's work is also important in another way.

His experiments showed that a critical, minimal plasma density is required for the Primer Fields to function.

It is also known from basic plasma physics, that in the real universe where the magnetic fields themselves are dynamically created by the flowing movement of electrically charged plasma particles, that the resulting geometry in free-flowing plasma, which is forged by the Lorentz force and its pinch effect, depends on a certain density of the plasma in motion. This means that two minimal thresholds are involved that must be superseded for the system to function, below which the primer fields become inactive. This means for the case of the Sun, that the Sun will become inactive when the plasma supply streams diminish below the system's minimal threshold point.

And this is precisely what stands before us in the near future.

Diminishment measured by NASA's Ulysses

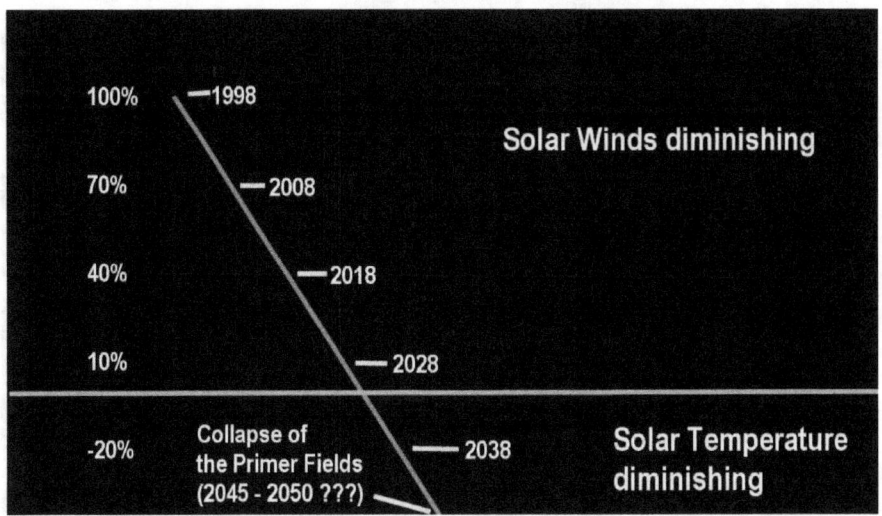

The current rate of diminishment of the solar activity, which was first measured by NASA's Ulysses spacecraft, between 1998 and 2008, which is still accelerating, suggests that the minimal threshold for the Sun going inactive may be crossed in the 2050s timeframe.

The long-held theory is wrong

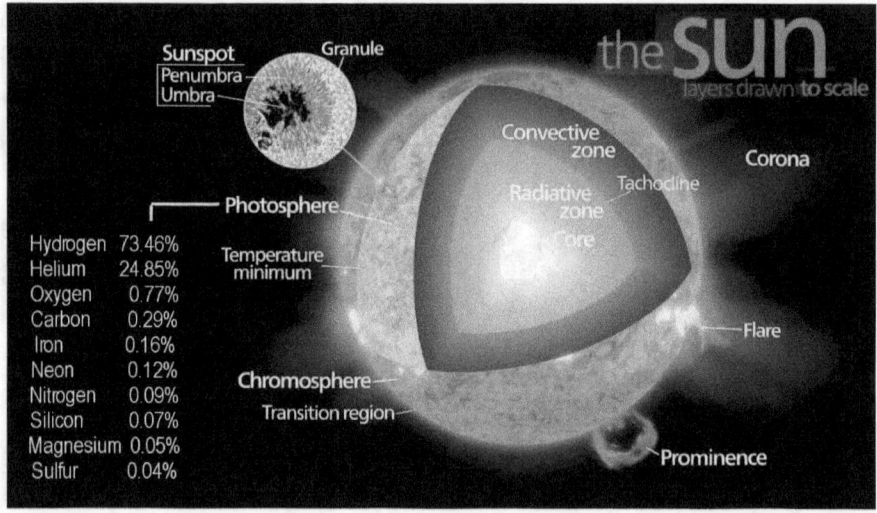

The more one explores the electric dynamics of solar physics, the more evidence comes to light that the long-held theory of the Sun being an internally heated gas sphere is wrong, that is deemed to be powered by a nuclear-fusion process occurring at tits core.

The plasma model of the Sun

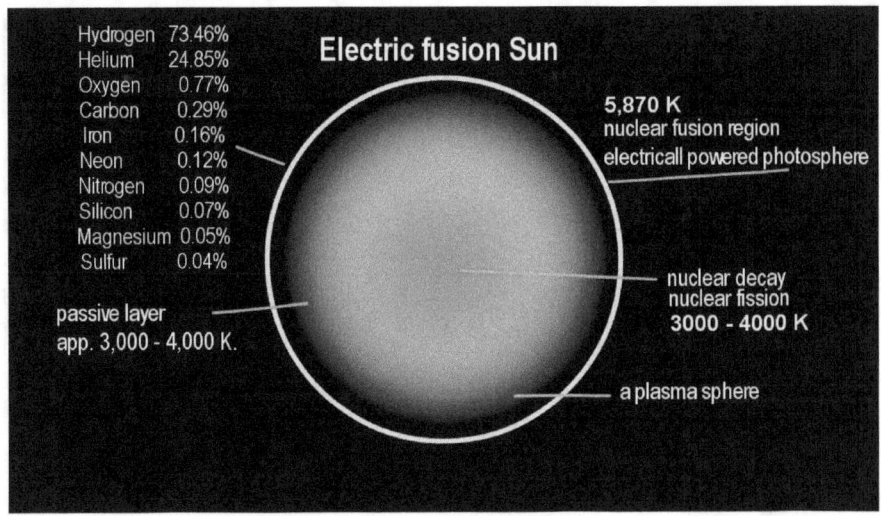

The evidence instead suggests that the Sun is essentially nothing more than a large plasma sphere with nuclear fusion processes occurring on its surface.

While the difference between the two models appears to be purely academic in nature, the opposite is actually the case. This is so, because the plasma model of the Sun, which all available evidence points to, renders our Sun as a variable star that is vulnerable to changing supply conditions, meaning that the Sun can go inactive for long periods.

When the minimal conditions are not met

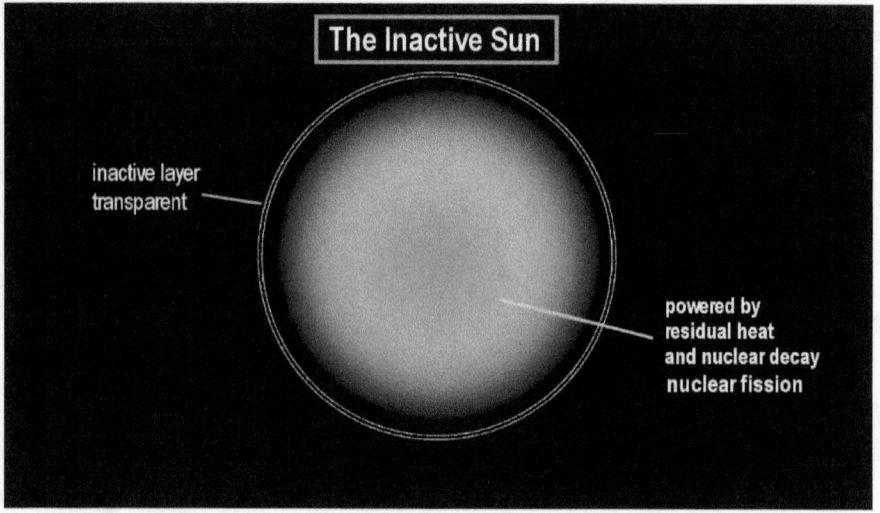

When the minimal conditions are not met, the surface nuclear-fusion reactions stop; the Sun will radiate only its stored-up 'residual' energy. It becomes darker and weaker then, whereby ice ages occur.

*The Plasma Sun model

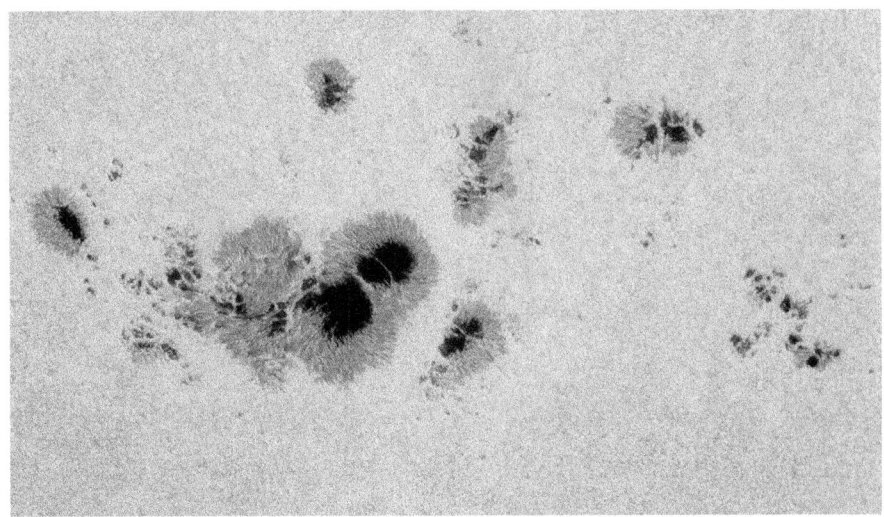

The view we get by looking through the sunspots, at what lays below the active surface, gives us an indication of what we can expect the entire Sun to be like in its inactive state.
The vulnerability of the Sun to the presently diminishing plasma-density conditions, appear to be so critical to human living that it becomes imperative for our very existence on this planet that the Plasma Sun model becomes understood as much as this is possible, and be acknowledged as the most critical factor in modern civilization until absolute conclusions can be reached, if this is indeed possible.

Our understanding of the Plasma Sun model

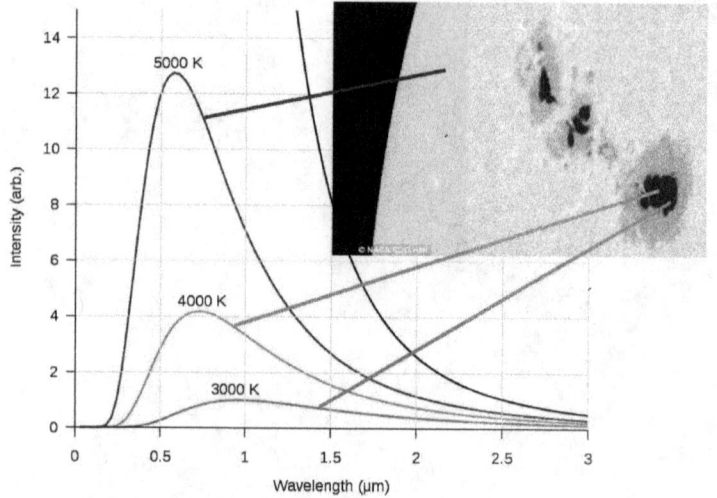

Nothing is more critical in modern civilization than our understanding of the Plasma Sun model, because from the moment on when the Primer Fields collapse and the Sun goes inactive, the Earth will receive potentially up to 70% less radiated energy from the Sun, which emits from this point on only stored-up energy, some nuclear-fission decay energy, and perhaps also some mild forms of energy from lesser plasma-interaction effects.

The Sun going inactive will disable agriculture

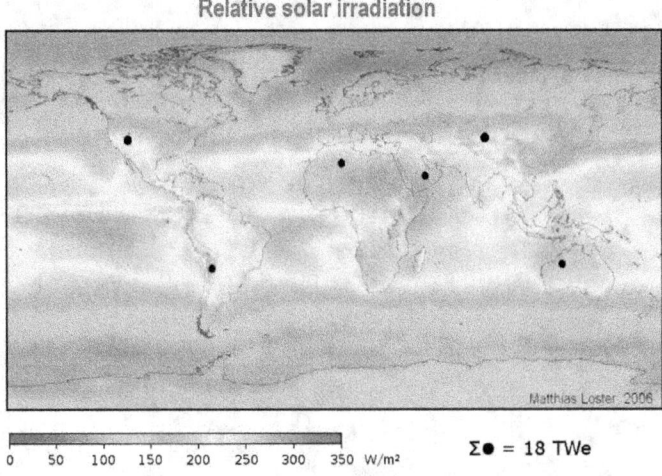

Relative solar irradiation

$\Sigma\bullet = 18$ TWe

It is obvious that an enormous reduction in solar energy input on earth will result from the Sun going inactive, which will disable agriculture almost over night above the 40 degree latitudes.

Between the Tropics of Capricorn and Cancer

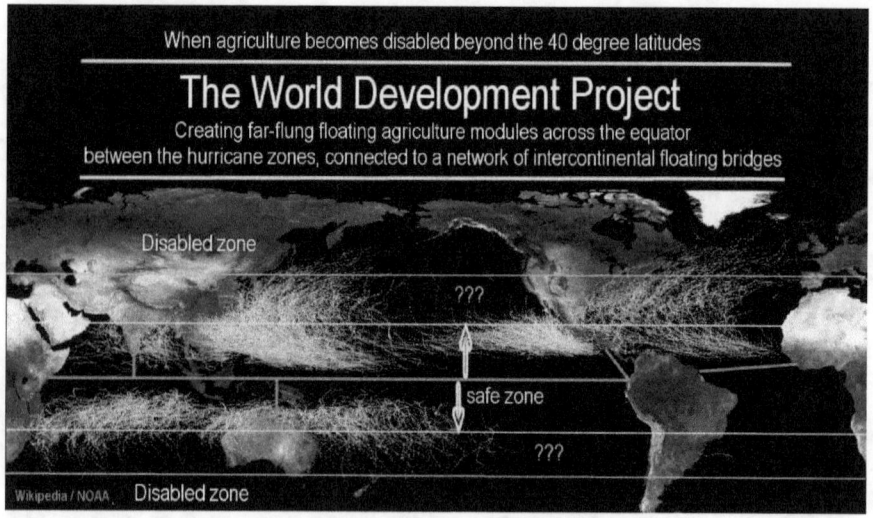

Agriculture might remain possible only across the narrow band between the Tropics of Capricorn and Cancer, centered on the equator, afloat on the sea between the hurricane zones, for the lack of sufficient land near the equator.

Higher latitudes, will become uninhabitable

Relative solar irradiation

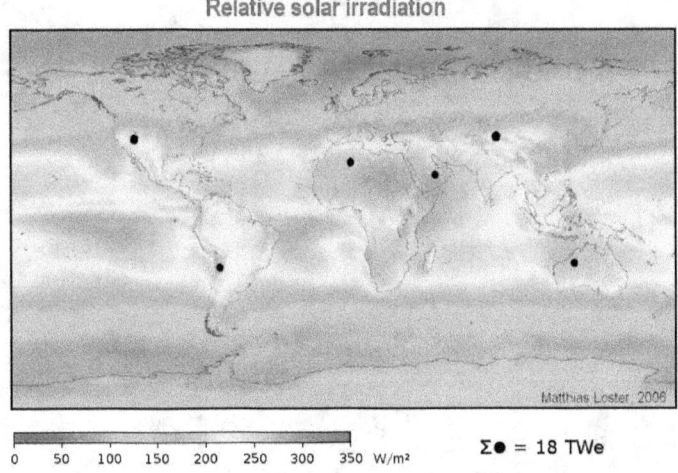

$\Sigma\bullet = 18$ TWe

All of the northern countries above the 40-degree latitude, due to the curvature of the surface of the Earth in the higher latitudes, will quickly become uninhabitable as the Ice Age begins rapidly under the inactive Sun.

For the affected nations to relocate

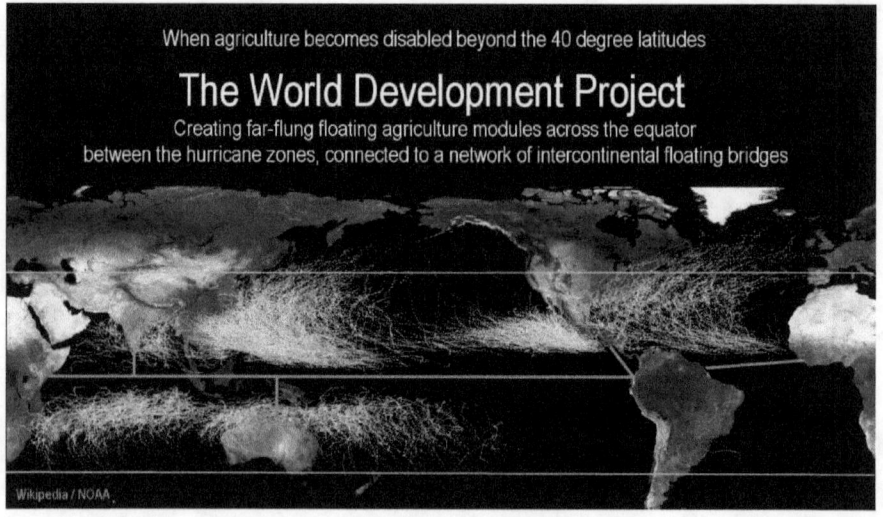

When agriculture becomes disabled beyond the 40 degree latitudes

The World Development Project

Creating far-flung floating agriculture modules across the equator
between the hurricane zones, connected to a network of intercontinental floating bridges

Wikipedia / NOAA

In order for the affected nations to relocate themselves, 6000 new cities will need to be built in the narrow band that remains liveable on earth. But will we do it?
The evidence for what would need to be done affects many areas of science.

A video presents 58 items of evidence

I have recently produced a video study as a summary overview that presents 58 items of evidence related to the dynamics by which the Sun could go inactive, potentially in the 2050s timeframe. The summary is amazingly substantial, though it lacks the details that required a number of underlying exploration videos to present. Here the Science Challenge begins.

Science challenge is not a technological challenge

The science challenge is not a technological challenge, for building the floating cities and agricultural modules. As large as these infrastructures are required to be, the physical task can be easily accomplished with fully automated, nuclear-powered, high temperature industrial processes. The technology exists. The materials exist. And the energy resources exist likewise. No science revolution is needed to implement what we already have at hand to be applied. The great science revolution is needed to inspire us to get the necessary type of building started.

More than 90% of humanity would die of starvation

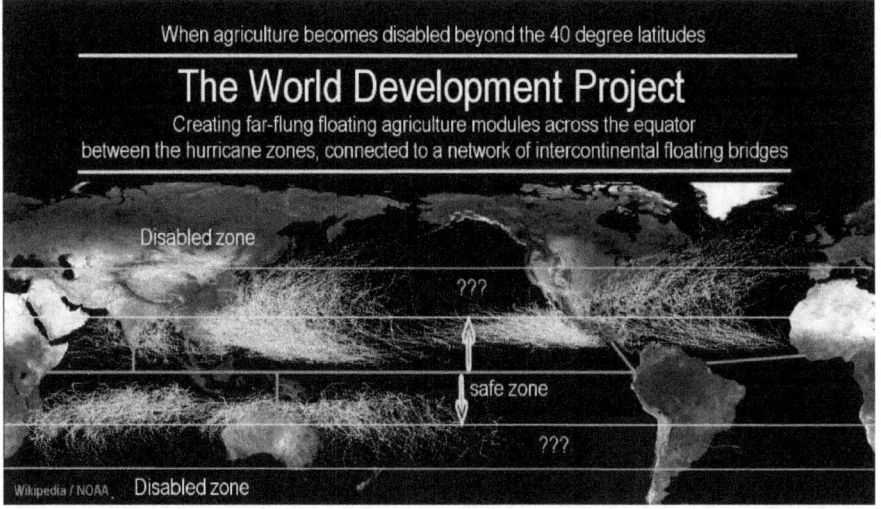

While it is widely acknowledged that the world's food supply is rooted almost exclusively in agriculture, in one form or another, it is rarely acknowledged that more than 90% of humanity would die of starvation if our existing food infrastructure became disabled, and no replacement agriculture had been built in protected regions to continue the food supply. The challenge to science is, to get society off the easy chair, and to get the 'spate into the ground' so to speak.

In comparison with the task of getting the world mobilized, the actual building of 20,000 kilometers of floating bridges and millions of acres of floating modules for agriculture and cities, and so forth, is easy.

To judge by the current rate of weakening in the solar system, it appears that we have enough time left to have the necessary construction tasks completed before the Sun goes inactive, which may occur in the 2050s. But will we do the work that needs to be done? Will we even consider it?

Most likely we will do nothing and commit thereby our children to an agonizing death.

The answer that we will give to ourselves in this regard, will be determined by our success in overcoming the science barriers that stand in the way, which have become a bulwark of shallow concepts and cultivated opinion, upheld by exotic theories of science 'epicycles' with no real evidence, fundamentally. It appears that we will need to heal science itself, first, before we will consider the needed answers truthfully.

Dad, why do you want to kill us children?

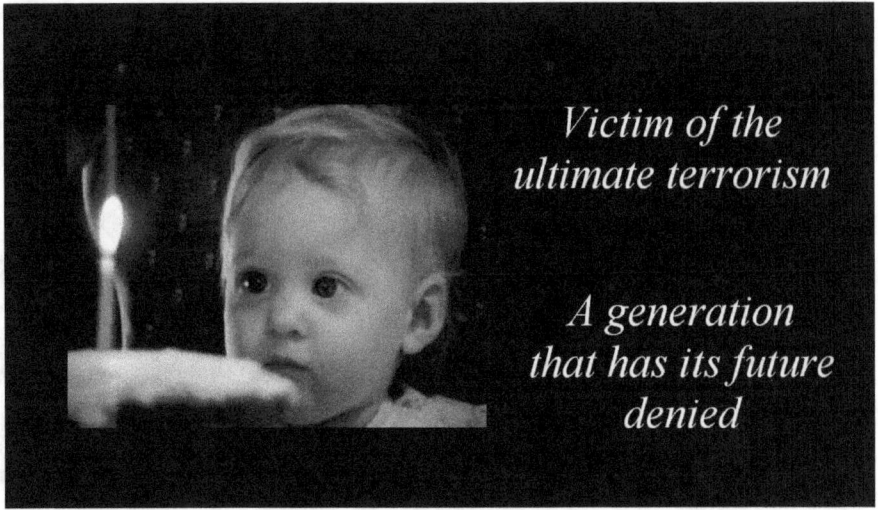

If we remain stuck in the current trap of shallow concepts and science epicycles, a child may ask in future years as it grows up, "Dad, why do you want to kill us children by not building a world for us that we can live in when the Sun goes dim?"

What will happen to us all when you are wrong?

"My child," the father might answer then, "don't be alarmed. The Sun will not go dim."

"But you don't know this for certain," the child might reply to such an argument, after it grew up enough to think for itself. "You can't say with certainty that the Sun will not go dim, because all the evidence points against your argument, indicating that the Sun 'will' go dim. And when it does, what happens then? What will happen to us all when you are wrong?"

What would a parent answer to that? A father might say, "my child, nobody believes the evidence that you speak of, to be what it appears to be."

Dad all your fancy theories that you cling to

"Oh, is this so?" the child may answer years later when the argument would come up again, and still nothing was done to prepare for the transition in the world that is to come. "I should say the same to you, Dad, about all your fancy theories that you cling to, to explain the impossible that isn't really happening, which you don't even pretend to have supporting physical evidence for, such as the theories of 'electron degeneracy' and the 'tunnelling effect' that you need to justify the internal fusion-sun theory, not to mention the 'dark energy' theorem that you precariously uphold to justify the dark umbra of the sunspots, and other exotic theories like them, such as the 'density-wave theory' that tells you what you observe with a telescope, when you look at the galaxies, doesn't actually exist."

What could a dad say speaking of justice?

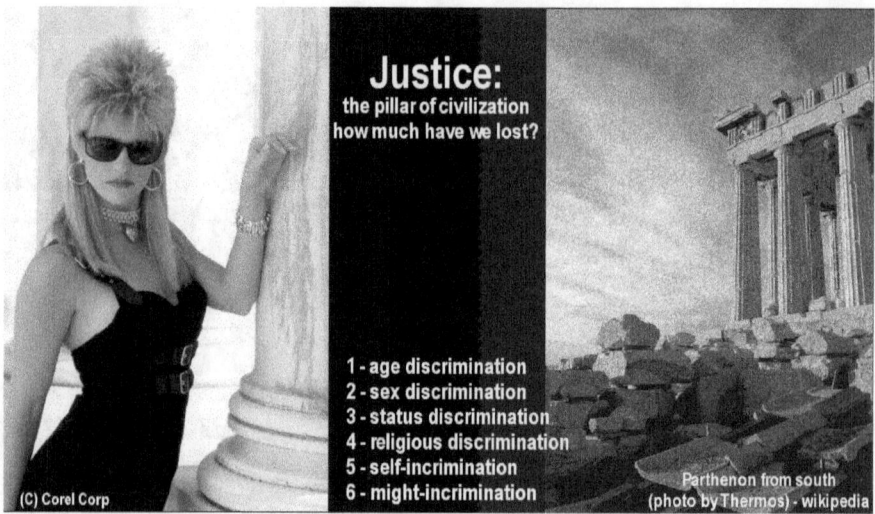

Justice:
the pillar of civilization
how much have we lost?

1 - age discrimination
2 - sex discrimination
3 - status discrimination
4 - religious discrimination
5 - self-incrimination
6 - might-incrimination

(C) Corel Corp

Parthenon from south
(photo by Thermos) - wikipedia

What could a dad say to that, if in later years he would face a grown woman, speaking of justice? Would he really believe in 'dark energy' and the like, and brush away his responsibility to be truthful, as easily as he did when he spoke with a child? Or would he begin to wonder if the sunspots are dark, because there is simply nothing behind the shiny surface, beneath the sunspots, as he had suspected, instead of the powerful nuclear fusion furnace that supposedly heats the Sun from the inside? Nor would he be able to explain to an educated adult why the Sun is far too light to be an atomic gas sphere of its size.

"We don't have all the answers," he might answer his grown daughter, uneasily. "We can only go with what we know."

You didn't get off your high chair

"That's a bad excuse," the adult daughter would say. "You didn't get off your high chair and made it your business to know the truth. If you had done this and had acted responsibly, and even if you had erred on the side of caution so that the ice age transition would never materialize that you would have upgraded the world for, then nothing would have been lost, regardless. We would have gained a richer world than we ever have dreamed of. Instead you remained glued to your damn chair and did nothing. Whatever was uncertain before compares as nothing now, with the certainty that we will all starve to death when the Sun goes dim, by which our agriculture fails that we depend on for our food. With this said, how much time do we have still left at this late date to build the infrastructures we need to save our existence? Probably too little. You have squandered too much of the precious opportunity we once had."

If you had only listened

"If you had only listened when I first raised the question to you, Dad, a long time ago, as a little girl," says the grown daughter to her dad. "It all seemed as like a dream then where nothing is real, and to you it all seemed academic in nature. I relied on you to tell me the truth, but you never did. You didn't have a good answer then, and never had, but you were not honest enough with yourself and with me, to say so. For this dishonesty you have committed us all to death, as have all the men and women whom you have failed to inspire with the truth. You have led the whole world to death, together with us, when the Sun goes dim, by doing nothing towards giving us a chance to have a future."

We still have a chance to avoid the condemnation

The Greatest Miracle on Earth
a human being
born in the image of God
reaching for spiritual light
to understand the universe
and itself

Fortunately, we are not at this point yet where such a conversation would unfold, where we would be condemned by our children. We still have a chance to avoid the condemnation, and treasure and protect what we have in each other.

In order that a conversation as the one illustrated here may never take place, I am presenting a summary of the physical evidence at hand, which a child might use in future years to indict us, if we let it come to that.

A wide scene of evidence, divided into three categories

I am presenting a wide scene of evidence, divided into three categories:
1 - Our Sun as an electric-fusion plasma star.
2 - Our Sun being variable and vulnerable.
3 - The solar cut-off timing

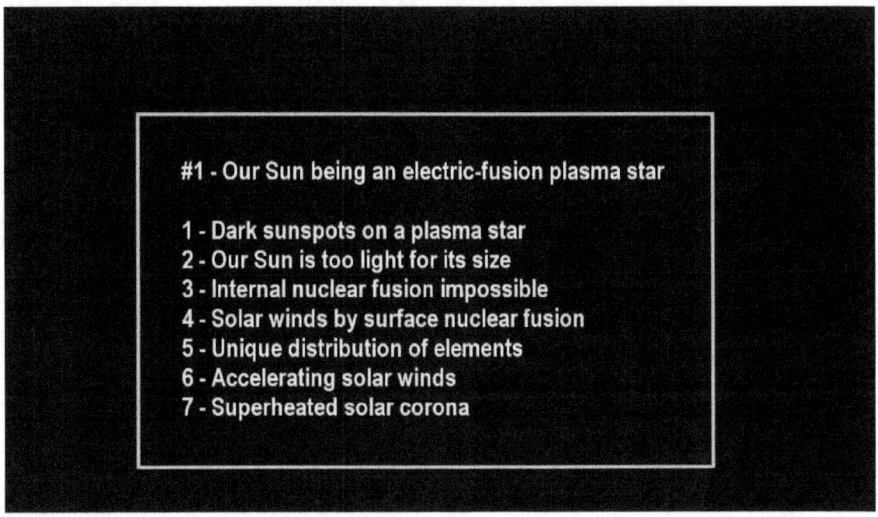

#1 - Our Sun being an electric-fusion plasma star

1 - Dark sunspots on a plasma star

2 - Our Sun is too light for its size

3 - Internal nuclear fusion impossible

4 - Solar winds by surface nuclear fusion

5 - Unique distribution of elements

6 - Accelerating solar winds

7 - Superheated solar corona

These are just a few points on this theme

The Sun is dark inside

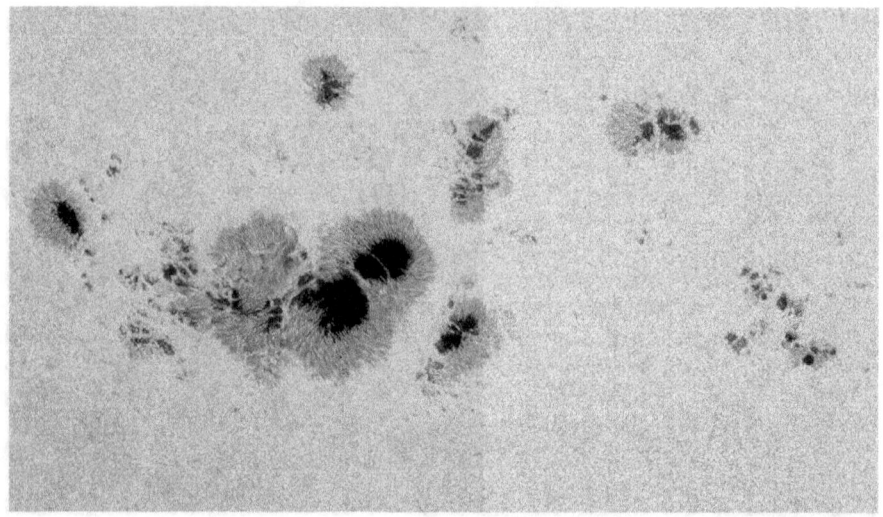

The plain evidence is that the Sun is dark inside, comparatively, when it is seen through the umbra of the sunspots. It is dark there, below the surface of the photosphere, because there is nothing there to be seen, with the Sun being a plasma star.

The Sun is impossible for a gas sphere

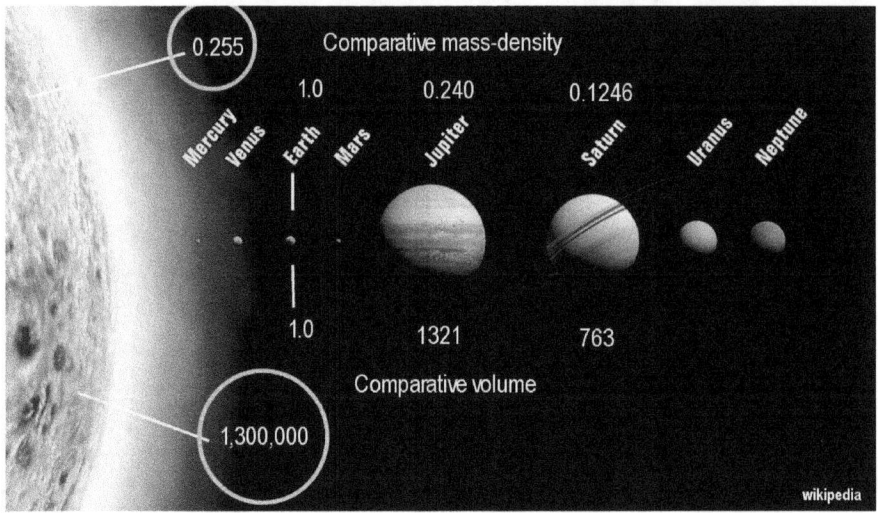

The Sun is a plasma star, because it is impossible for a gas sphere of the size of the Sun to exist. Its gas atoms would be crushed in the core by its massive weight. In a plasma sphere no such problems can exist. Its mass-density is determined by electric repulsion.

The fusion process would blow the Sun out

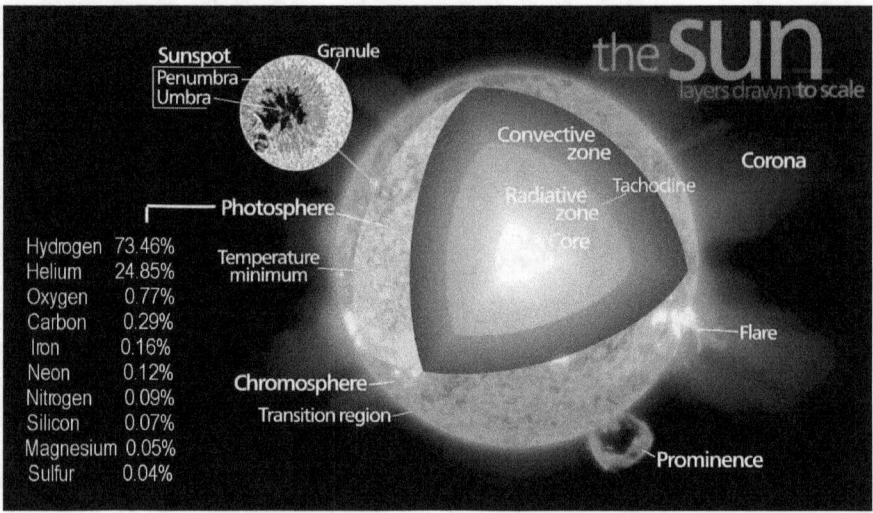

Nor is it possible for the Sun to operate a nuclear fusion process at its core, where hydrogen atoms are said to be fused into helium atoms. The fusion process would blow the Sun out. The core would become clogged with its fusion product that dilutes the fusion fuel, by which the reactions stop.

The Joint European Tokomak

On Earth, the most advanced nuclear fusion reactor, the Joint European Tokomak, blows itself out after a burn-time of slightly less than a single second, for this vary reason. The Sun can't operate on this platform.

The Sun can only operate as a plasma sphere

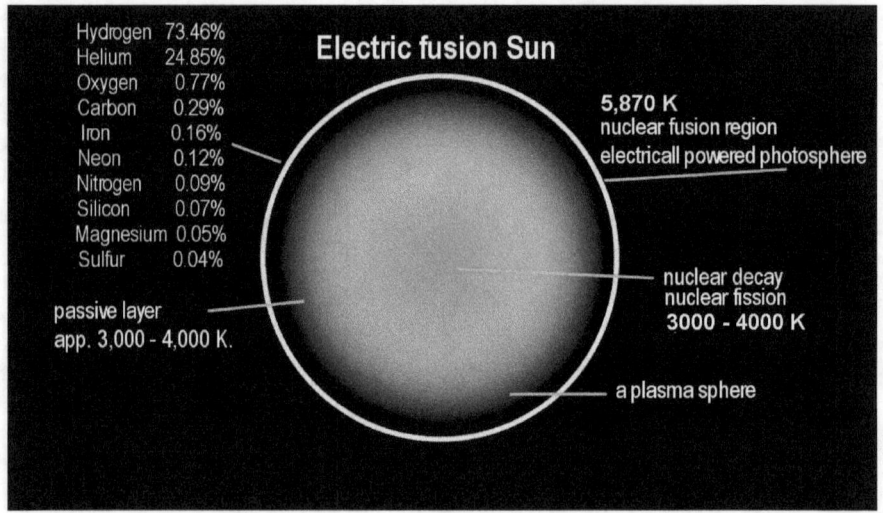

The Sun can only operate as a plasma sphere with electric nuclear fusion occurring on its surface, from where the fusion products are carried away by the solar winds.

That the Sun is powered by efficient nuclear fusion

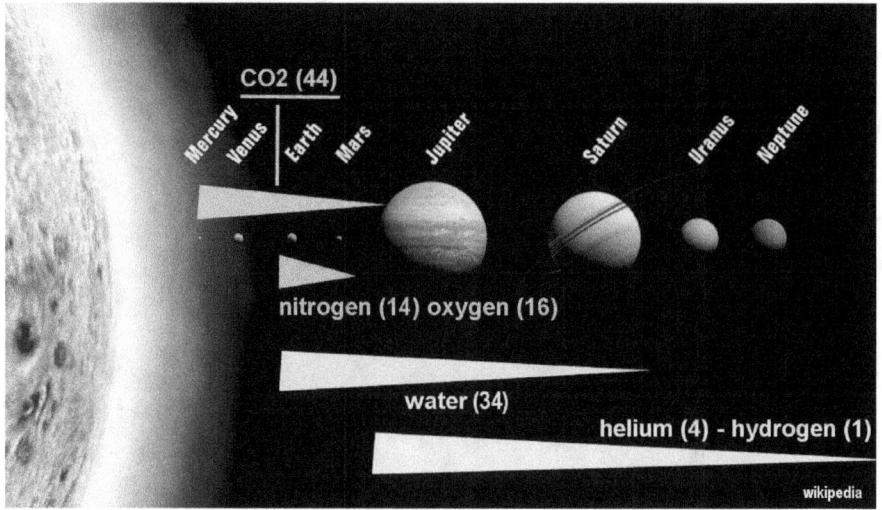

That the Sun is powered by efficient nuclear fusion occurring on its surface, in which all the elements in the periodic table are synthesized, is evident by the unique distribution pattern of the elements across the planets. All the inner planets are made up of heavy elements for this reason, which are the first to be attracted out of the solar winds, while hydrogen, as the lightest element, gets carried as far away as Pluto, this wouldn't happen if the Sun was not the central synthesizing source of all the elements, which it is, since all the basic elements have been detected in the solar atmosphere.

That the Sun is an electric star is further evident

© Miloslav Druckmuller/Barcroft
http://www.zam.fme.vutbr.cz/~druck/Eclipse/
- an example of the amazing solar eclipse photography
of Miloslav Druckmueller

That the Sun is an electric star is further evident by the existence of the solar wind that is accelerated away from the Sun to speeds of 800 kilometers per second. What we see here, is the natural by-product of the dynamics that enable electric nuclear fusion to occur. If the Sun was internally powered, solar winds would not occur.

The superheated corona

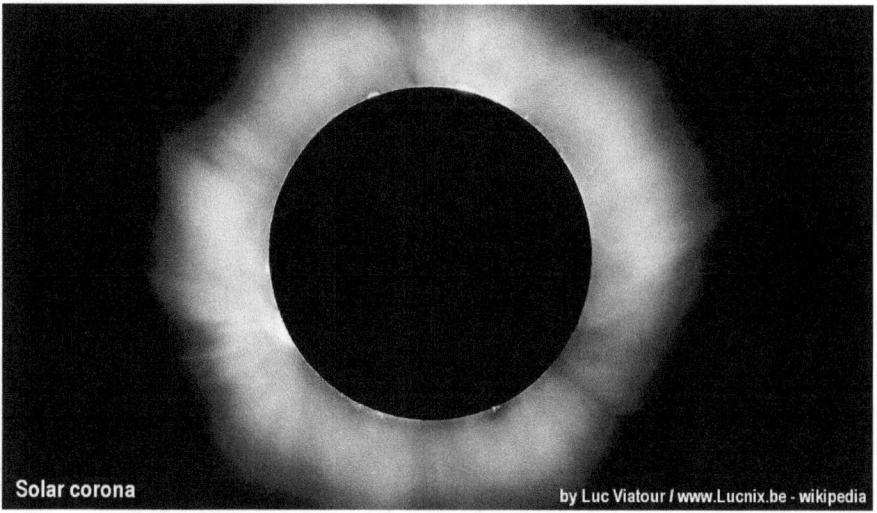

Solar corona

The superheated corona at hundreds of times of the surface temperature of the Sun, would likewise not occur if the Sun was heated from within.

The superhot corona is possible and natural

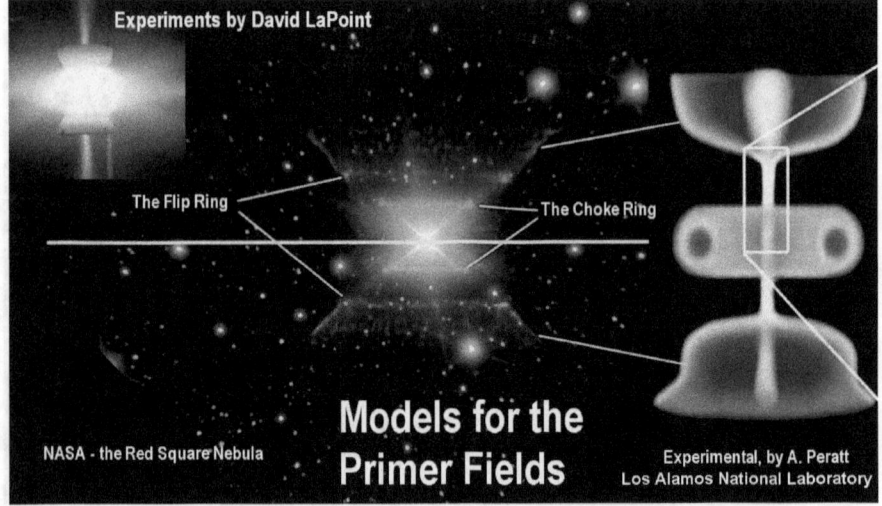

The superhot corona is possible and natural only for a Sun that has a large sphere of highly condensed plasma focused on it, by its Primer Fields. No magic is involved here.

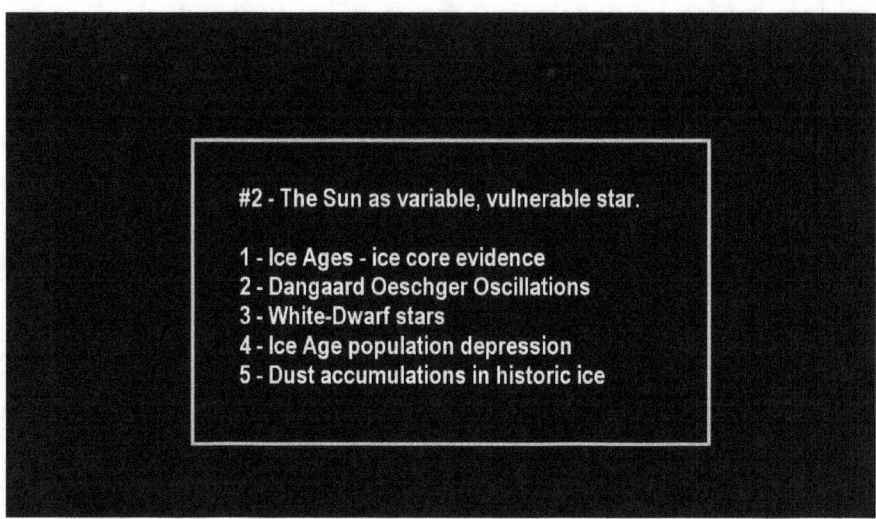

#2 - The Sun as a variable, vulnerable star.

1 - Ice Ages - ice core evidence

2 - Dansgaard Oeschger Oscillations

3 - White-Dwarf stars

4 - Ice Age population depression

5 - Dust accumulations in historic ice

The evidence is plain

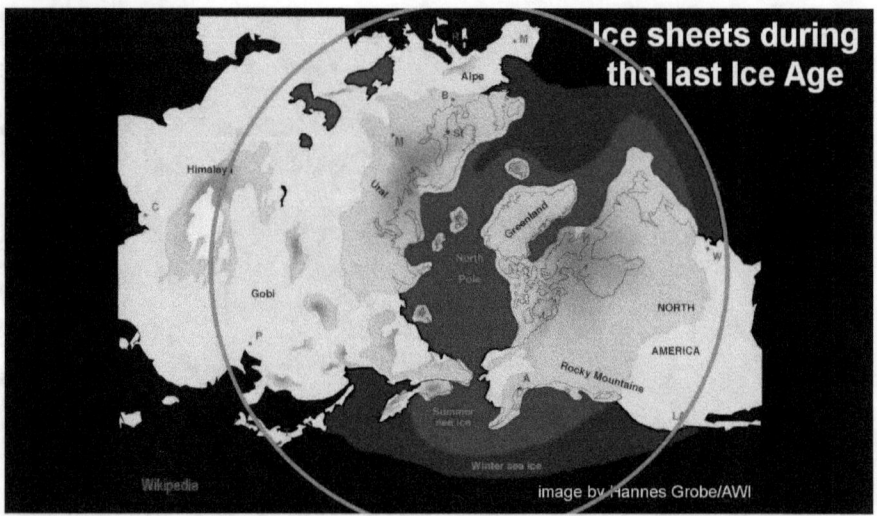

The evidence is plain. Ice Ages would not be possible if the Sun was an invariable constant, in the climate 'theater.' We would never have had Ice Ages. Still we had them.

Ice Ages have occurred as long as humanity has existed

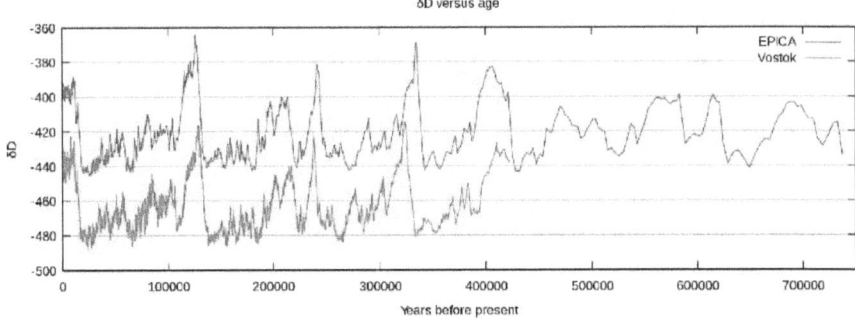

Ample evidence exists that Ice Ages have occurred almost for as long as humanity has existed, spanning more than 2 million years. No actual theory exists for the known Ice Ages to have occurred under a constant Sun, which renders the Ice Ages that we have records of, to be the phenomena of our Sun going inactive, from which Ice Ages result.

Dansgaard Oeschger Oscillations

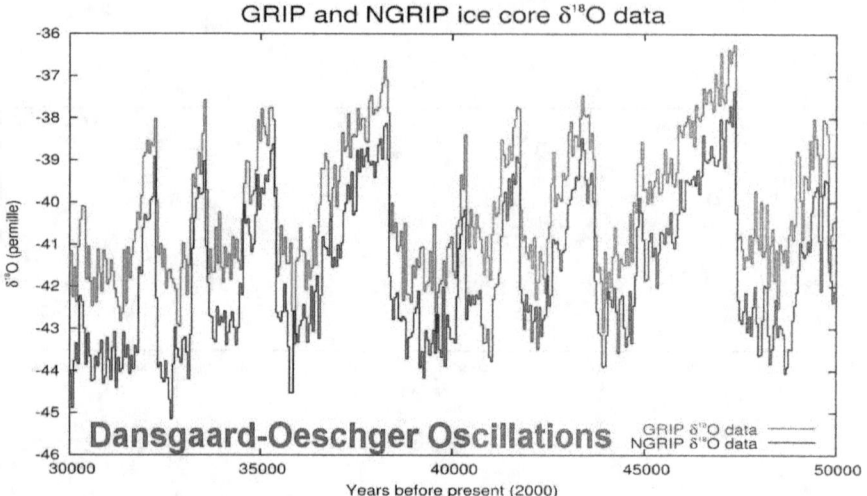

The Dansgaard Oeschger Oscillations that have occurred all the way through the last Ice Age, which we have ice core records of from two sites of Greenland, are not possible under a constant Sun, but are the natural result of the Sun being inactive all the way through the last Ice Age, with the exception of 25 brief occurrences when the Sun became active for a few decades at intervals of typically 1470 years, and then cooled down again.

Our active Sun, a rare anomaly

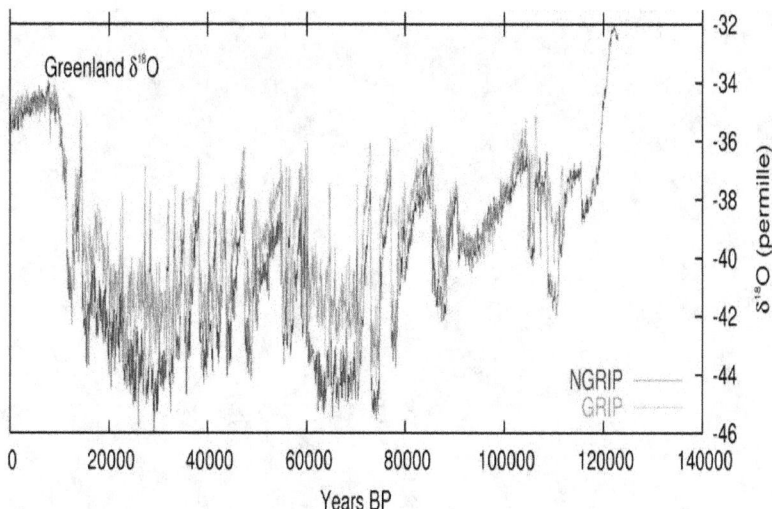

These enormous, rapid oscillations between deep glacial conditions and near interglacial conditions are only possible with the Sun oscillating between its active and inactive states, with the inactive state being the normal state for our Sun for 85% of the last 2 million years. In this context, the Dansgaard Oeschger Oscillations render our currently continuously active Sun, a relatively rare anomaly, that we are about to get back to in a few decades.

A large number of white dwarf stars

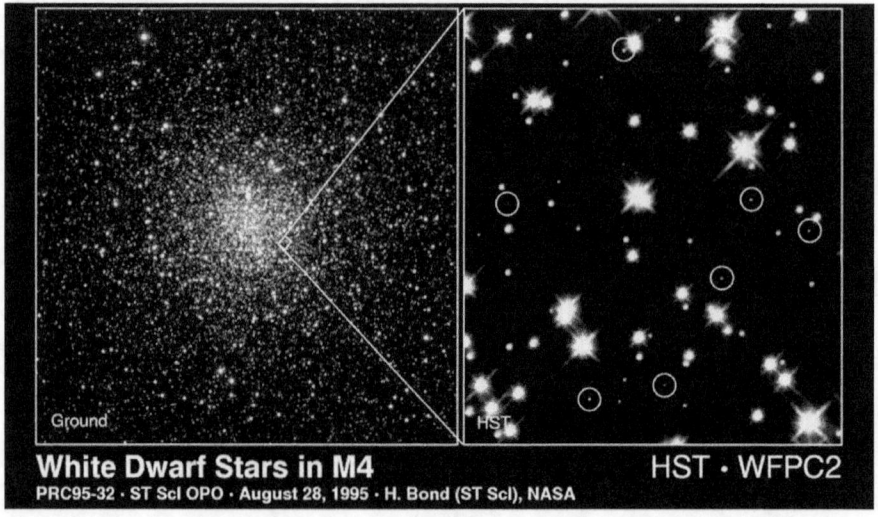

White Dwarf Stars in M4   HST · WFPC2
PRC95-32 · ST ScI OPO · August 28, 1995 · H. Bond (ST ScI), NASA

That the Sun is not alone in assuming an inactive state may be evident the in the existence of a large number of white dwarf stars. This is the type of star that one would expect to see for our Sun, after it goes inactive for long periods, where it remains aglow only by slow nuclear decay of synthesized atomic material that the Sun had retained with its gravity, from its active times, which slowly drift into the interior to be crushed by the gravitational and electric forces in its core.

During the initial ice-age transition

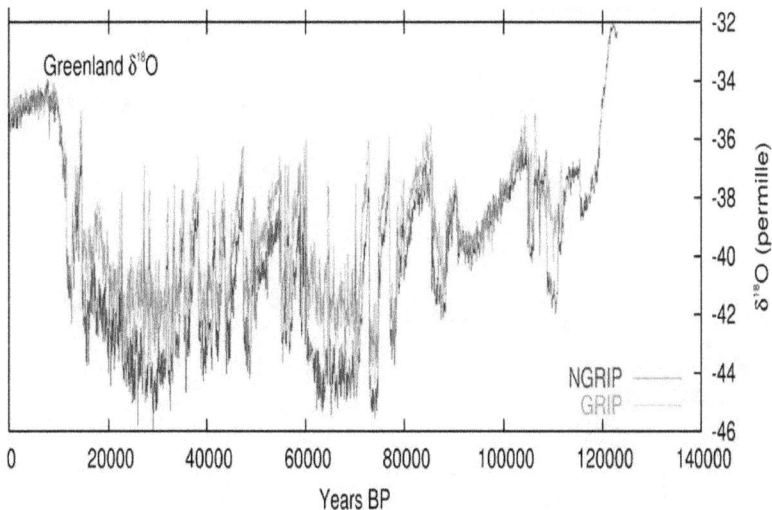

During the initial ice-age transition, as 120,000 years ago, towards a white dwarf star state, our Sun may have gone through several intermediate stages, such as the so-called red-dwarf stage. We really don't know what precisely the sequence would have been. The last Ice Age transition from an interglacial period had occurred 120,000 years ago. No records were kept.

When the global climate suddenly gets 40 times colder

Nor do we really know what the type of huge transition means, that is indicated here, when the global climate suddenly gets 40 times colder than the Little Ice Age in 1600s has been, which is barely remembered, but is the only reference that we have available for comparison.

Severely depressed living conditions

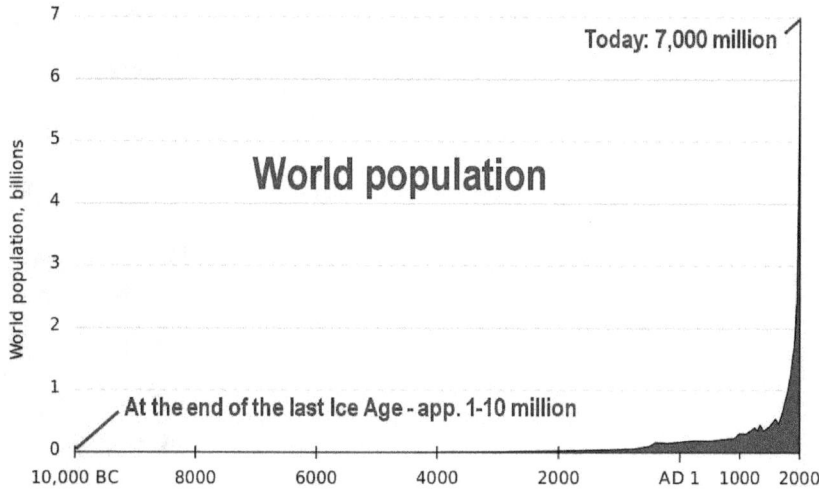

Archaeological evidence speaks of severely depressed living conditions occurring all the way through the Ice Ages. The conditions were evidently so harsh that at the end of the last Ice Age the entire world population had shrunk to a mere 1 to 10 million people. This was the end result, at this time, of more than 2 million years of human development.

The minuscule population level, following the glaciation period, speaks volumes about the harsh conditions that evidently existed in our world under a mostly inactive Sun, which we call the Ice Ages for the resulting effects.

The deep population depression adds one more item of proof that the Sun can and will go inactive, and has done so in the past.

Dust accumulation in the ice of Antarctica

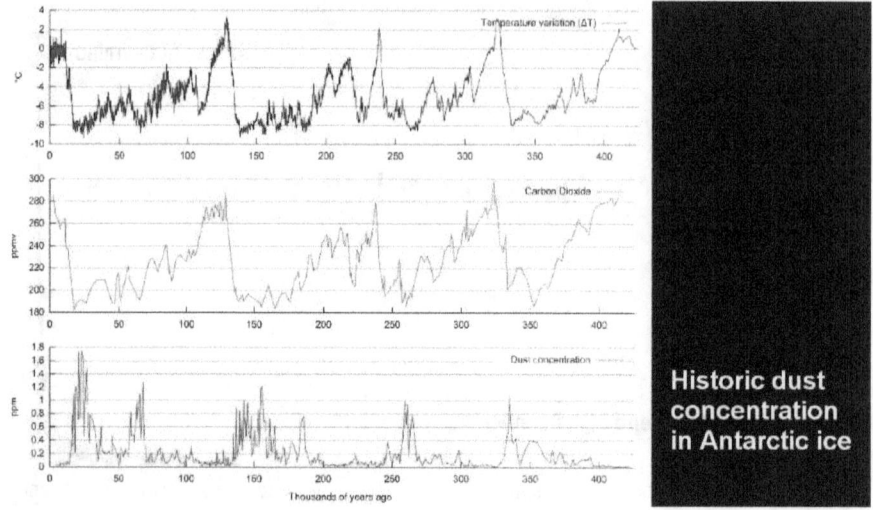

Historic dust concentration in Antarctic ice

Another item of evidence for the Sun being mostly inactive during the Ice Ages, is found in the form of dust accumulation in the ice of Antarctica. The dust accumulations typically begin half-way through the glaciation periods, and extend to the start of the next interglacial period. This type of phenomenon is natural for the resulting condition when the main Primer Fields for the Sun, which normally stabilize the orbits of the planets, cease to exist for long periods in which the Sun remains inactive.

Under inactive conditions

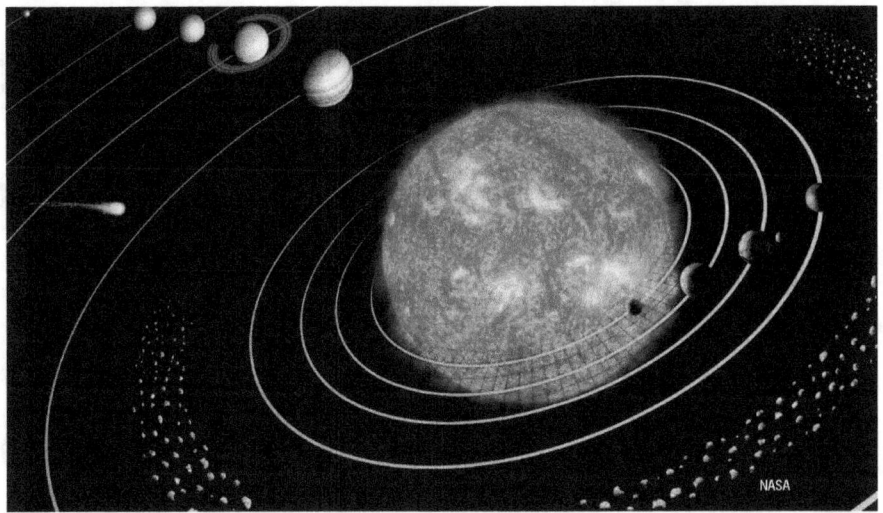

Under such 'default' conditions, or inactive conditions, the orbits of the asteroids in the asteroid belts, and in near-Earth space, invariably begin to decay. They become disorganized. Asteroids have a large surface to mass ratio and are more readily slowed by external influences. Eventually they begin to bombard the atmosphere of the Earth where they disintegrate into dust. The asteroid intrusion has spanned in previous times more than 40,000 years. In each case it stopped abruptly when the interglacial period began and the orbits became regulated again.

#3 - The solar cut-off is near

1 - Diminishing solar wind pressure
2 - Diminishing of underlying solar magnetic field
3 - Increased Galactic Cosmic Ray flux
4 - Earth's magnetic pole drift
5 - Earth magnetic field strength diminishing
6 - Diminishing sunspots and solar flairs

#3 - The solar cut-off is near

1 - Diminishing solar wind pressure

2 - Diminishing of underlying solar magnetic field

3 - Increased Galactic Cosmic Ray flux

4 - Earth's magnetic pole drift

5 - Earth magnetic field strength diminishing

6 - Diminishing sunspots and solar flairs

Ulysses spacecraft saw an amazing decline

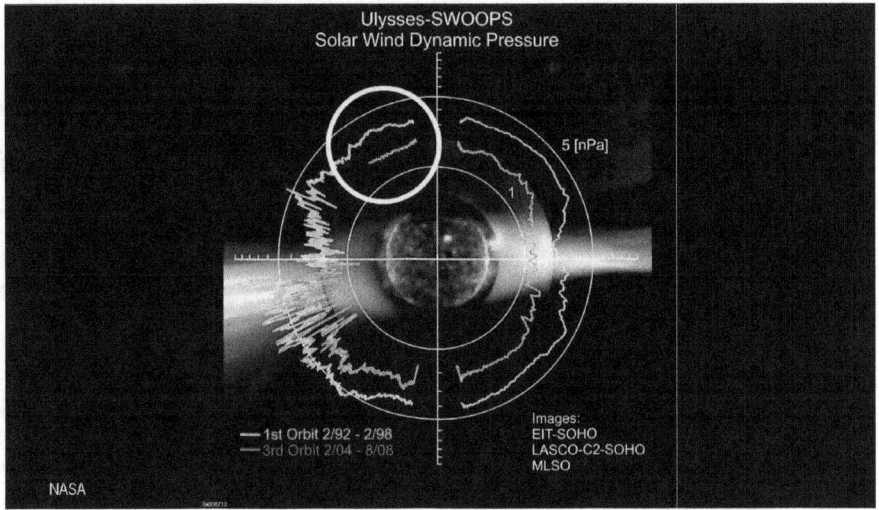

NASA's Ulysses spacecraft saw an amazing decline of the solar wind pressure, by 30% over the span of a decade. It saw a new trend unfolding that is still continuing.

The solar wind is a by-product

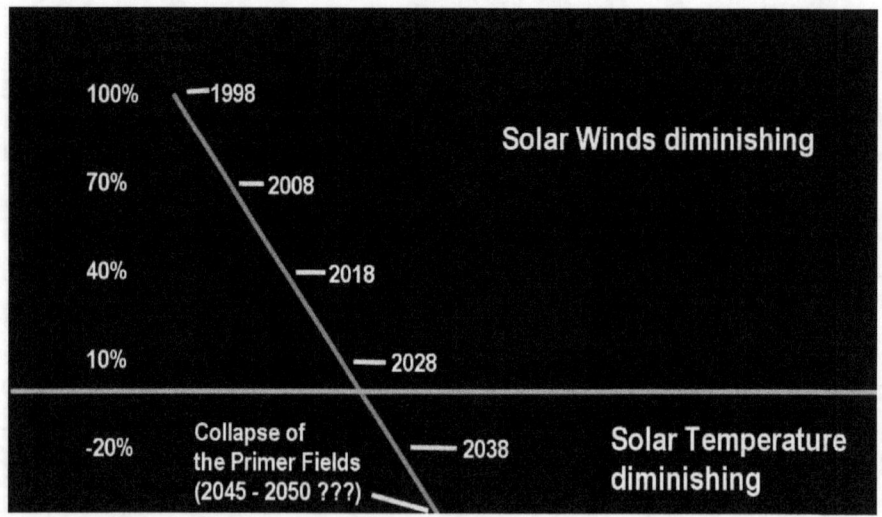

If the trend is projected forward in a linear manner, the solar wind will diminish to zero in the 2030s timeframe. Since the solar wind is a by-product of the regulating system that keeps the Sun at a relatively constant level, the solar-wind cut-off point would mark the beginning of the weakening of the Sun's energy production. The linear continuing of the weakening trend thereafter, would likely lead to the collapse of the Primer Fields in the 2050s timeframe.

A 20% increase of Galactic Cosmic Ray flux

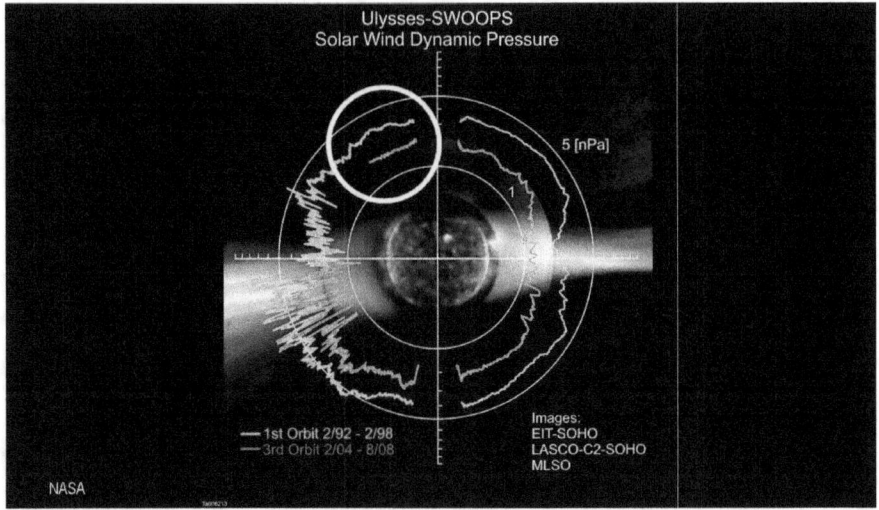

Coincident with the measured weakening of the solar wind pressure, the Ulysses spacecraft had also measured a 30% reduction of the strength of the Sun's underlying magnetic field, and a 20% increase of Galactic Cosmic Ray flux penetrating the barrier provided by the solar heliosphere that weakens with the weaker solar-wind pressure.

A wide variety of meteorological effects

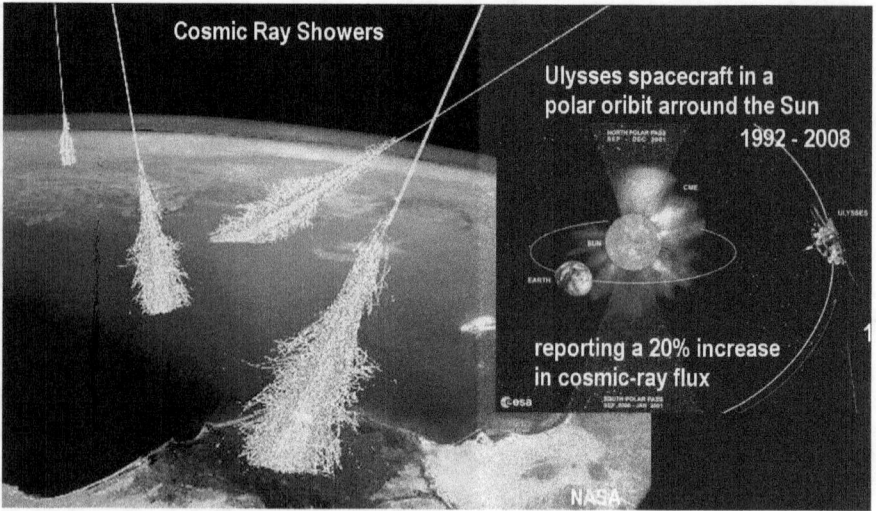

The weakening of the solar wind has become evident all over the world in the form of a wide variety of meteorological effects.

Increased cloudiness

ISS-34 - Stratocumulus clouds

Increased cloudiness is one of the effects. The increased white surface reflects more of the solar radiation back into space, which becomes lost to the heat budget on Earth, with cooling effects in the higher regions of the troposphere.

The CLOUD experiment

CERN - CLOUD project - Jasper Kirkby

The CLOUD experiment at the European Organization for Nuclear Research, one of the world's largest and most respected centres for scientific research, has demonstrated in principle that cosmic-ray interaction with the atmosphere is enormously effective in increasing cloud nucleation.

The measured effects

The measured effects went straight up and off the chart.

The resulting colder atmosphere

The resulting colder atmosphere reduces the distance of the water transport by the clouds, which results in increased drought condition and loss of agriculture.

The increased cloud nucleation

The increased cloud nucleation, of course, reduces the water vapor density, which is the main contributor to the moderating greenhouse effect of the atmosphere, up to 97% of it. With the moderating effect diminishing, we experience larger differences in temperature fluctuations, including colder winters, hotter summers, and ironically also arctic warming during the arctic summers, as the result of the diminished moderating greenhouse.

All of these effects are already experienced worldwide, while the continuing trend of the electric weakening of the solar dynamics will cause these experienced 'fringe' effects to become increasingly more severe over the next 30 years.

The reduced deflection of the Earth's magnetic field

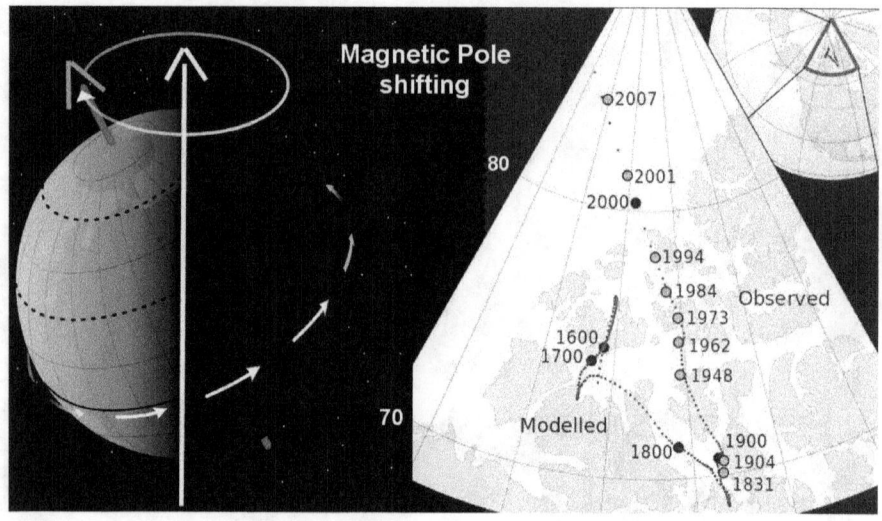

The continued weakening of the plasma system that is powering our Sun, is also directly measurable on the surface of the Earth. It is measured in the form of the reduced deflection of the Earth's magnetic field. Without external influences, the Earth's magnetic field would be aligned with its spin axis. With strong external influences, the magnetic poles become shifted away from the geographic poles by 23 degrees according to the inclination of the spin axis. With the external influence rapidly diminishing, the magnetic pole is drifting back towards its 'normal' spin-axis orientation.

The electric weakening is accelerating

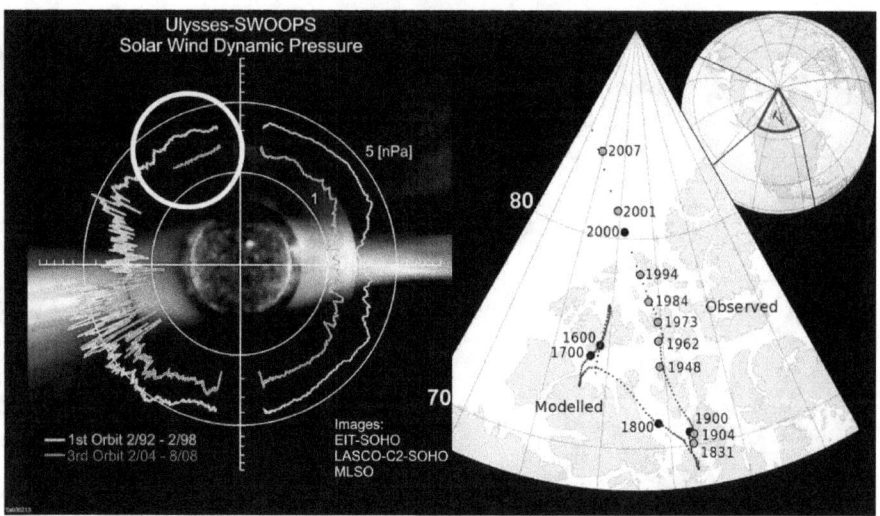

While other factors are involved in the observed phenomenon, the indicated pattern seems to suggest the electric weakening of the solar dynamics system is accelerating, instead of remaining linear. The magnetic pole drift is even more ominous when one considers that the Earth magnetic field itself is weakening, and is said to be at its weakest state since records were kept.

Not limited to the Earth alone

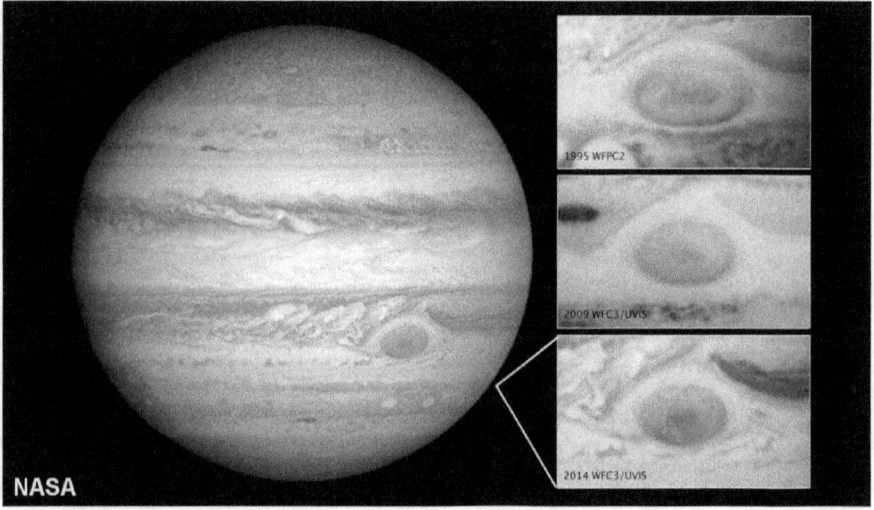

The effects of the diminished electric solar dynamics are not limited to the Earth alone, but are also evident on the planets. One observed effect is the shrinking of the great read spot on Jupiter that became visibly smaller since the weakening trend began.

The solar flair index that is becoming noticeably weaker

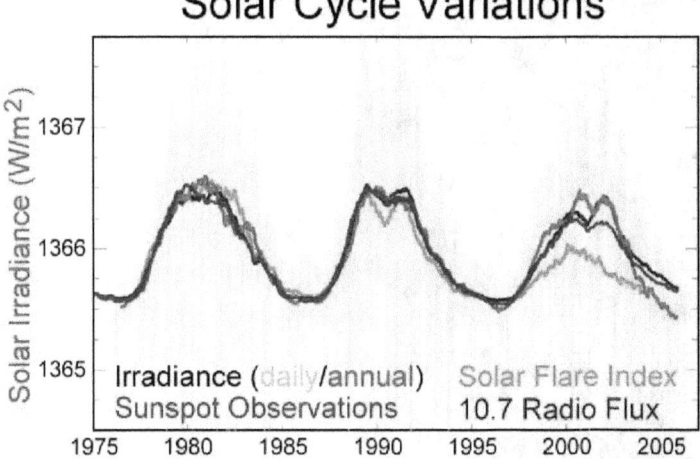

Solar Cycle Variations

Another item of evidence that a major weakening trend is in progress, is evident in the solar flair index that is becoming noticeably weaker, since the weakening of the electric solar dynamics began.

The sunspots are getting weaker

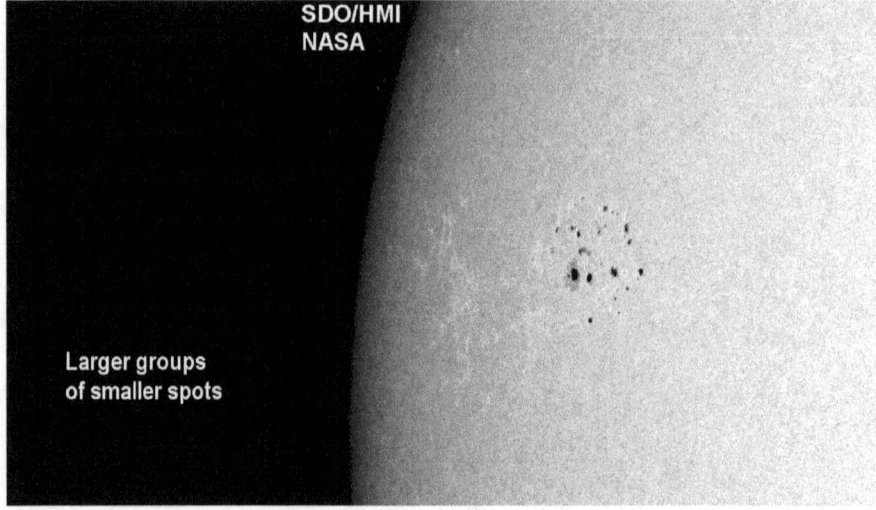

Even the sunspots themselves are getting gradually weaker. The huge spots that blow out massive amounts of internal pressure, are getting fewer, while smaller spots result, that tend to be surrounded by a sea of little spots.

It is impossible to forecast the exact time

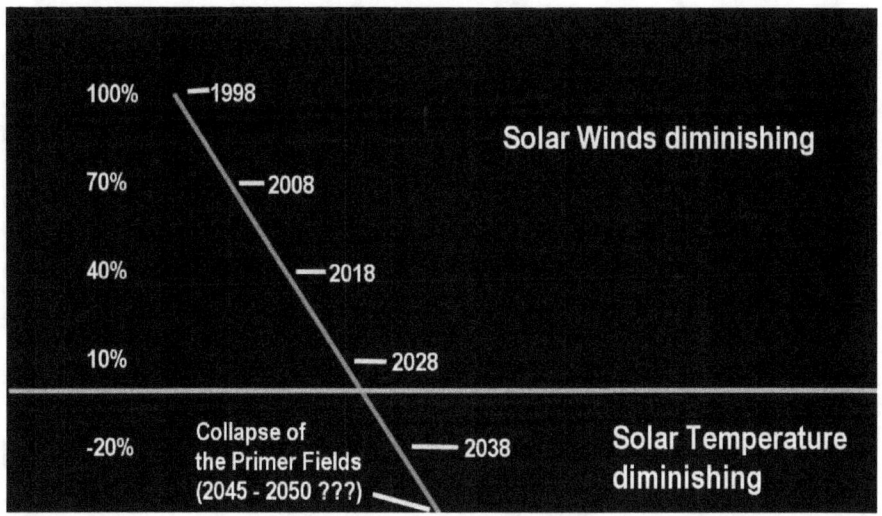

It is impossible, of course, to forecast the exact time when the fragile solar dynamics system goes inactive. Nor is this necessary. We know the trends. We understand the principles involved. We know the consequences. And we know what to do to reorganize our world to be able to live with the consequences. But will we do it?

It is far easier to remain asleep in our chair

Sure, it is far easier to remain asleep in our chair with the comfortable illusion that the world will grind on forever, as it has since civilization began, especially so as the illusions are supported by scientific assumptions that would make us believe that nothing needs to be done, as the world will grind on forever. It is this comfortable indifference based on illusions for which no real physical evidence exists, that renders the current Ice Age Challenge the greatest Science Challenge of our time.

Not located in developing the physical means

The Science Challenge is not located in developing the physical means for building the infrastructures for the continued existence of humanity in an Ice Age world. The physical solutions are not difficult to implement in principle, with automated industrial production. But will we do it?

*The challenge to science is to get itself out of the easy chair

The easy chair
is free

death by default

The challenge to science is to get itself and humanity out of the easy chair, and this in time before the Ice Age transition begins, with the Sun going inactive.

An intimate challenge to ourselves

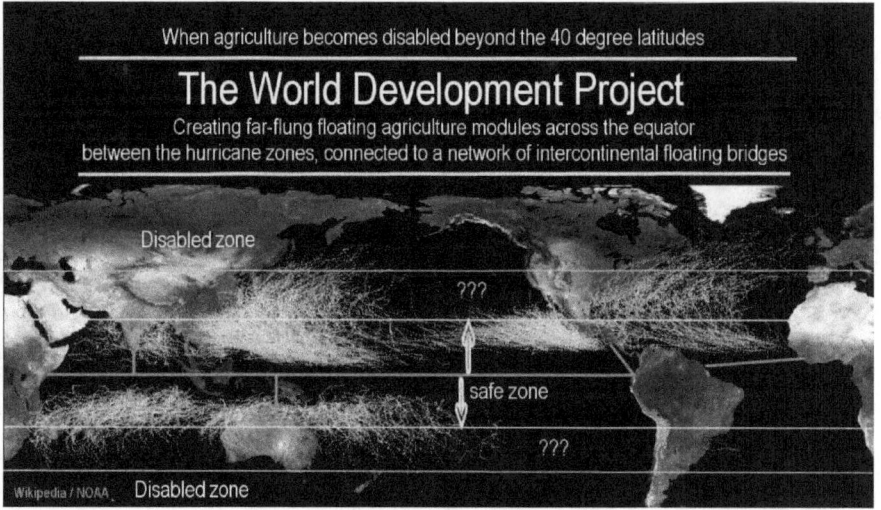

When agriculture becomes disabled beyond the 40 degree latitudes

The World Development Project
Creating far-flung floating agriculture modules across the equator
between the hurricane zones, connected to a network of intercontinental floating bridges

We have the materials, technologies, and energy resources on hand to build what needs to be built to assure our continued food supply. But will we use these resources? This question makes the Science Challenge also an intimate challenge to ourselves that challenges our attitude towards our children.

We are that future generation

It is tempting to stay glued to our chair, and to say that the challenge is not ours, that it is in the court of future generations, thousands of years distant. However, the evidence seems to indicate that we are that future generation, so that the ball is in our court.

We are fast running out of energy resources

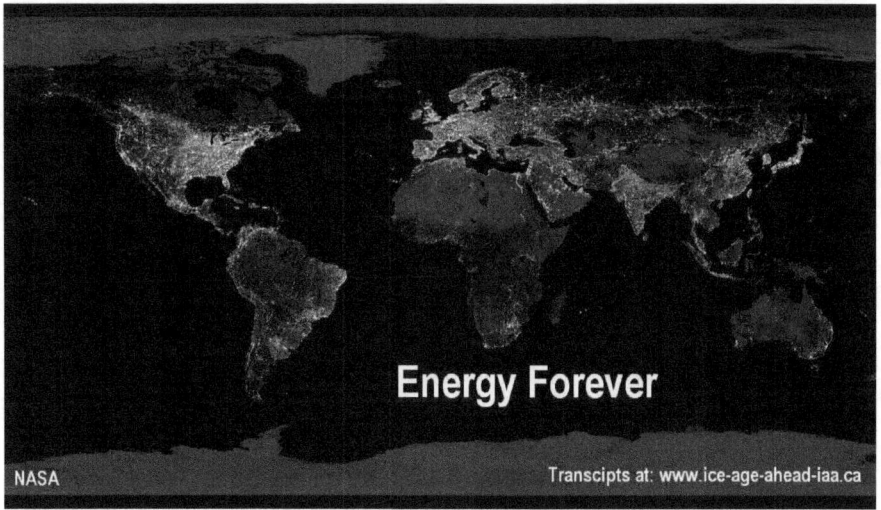

In fact, we should celebrate that the ball is in our court, because we need the science breakout to what is real, because we are fast running out of energy resources, while ignoring the anti-entropic energy resources that we have within the electrically powered universe and solar system.

Our entropic energy resources have already been largely depleted

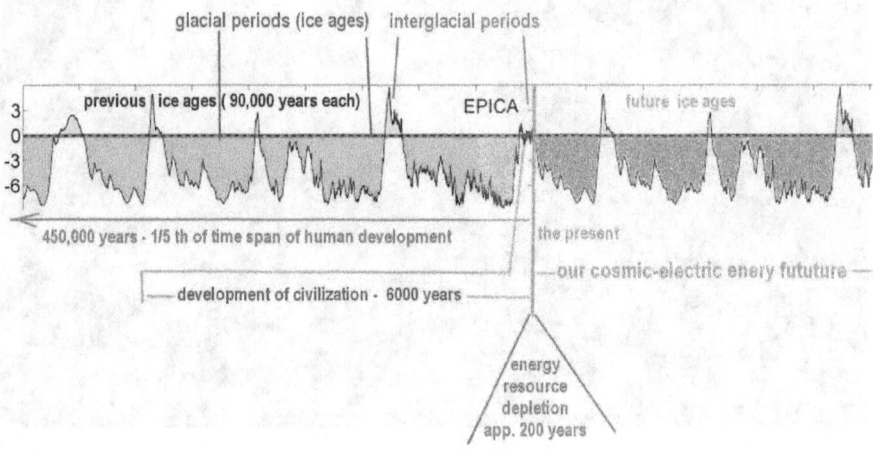

The development of our civilization has been relatively short, spanning roughly 6000 years. In the last 200 years of it, we began the large-scale energy use. During this brief span, our entropic energy resources have already been largely depleted. Some say we have only 60 years of oil left at the current rate of consumption, and maybe 400 years for nuclear energy. Then what? On the larger scene, with Ice Ages lasting for 90,000 years, it does not really matter whether we run out of oil in 60 years or in 1000 years, we need an anti-entropic energy source to have an endless future.

We have no real option left

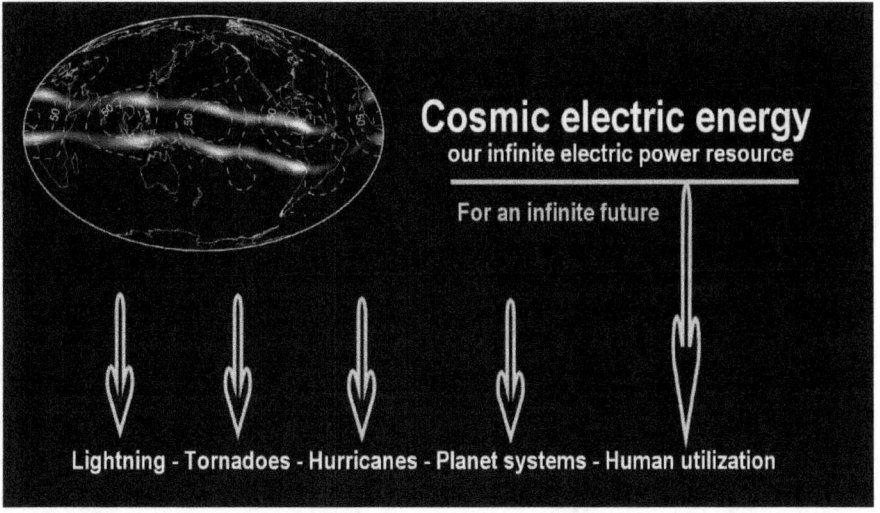

We have no real option left, even now, not to develop the cosmic electric energy resources that can never be depleted, which promise to be sufficient to meet our relatively minuscule human needs, for all times to come, even in solar inactive times when the interstellar plasma streams have a wider focus. The Science Challenge definitely includes, to get us moving into the cosmic-energy direction as well.

As one people of a single, universal, humanity

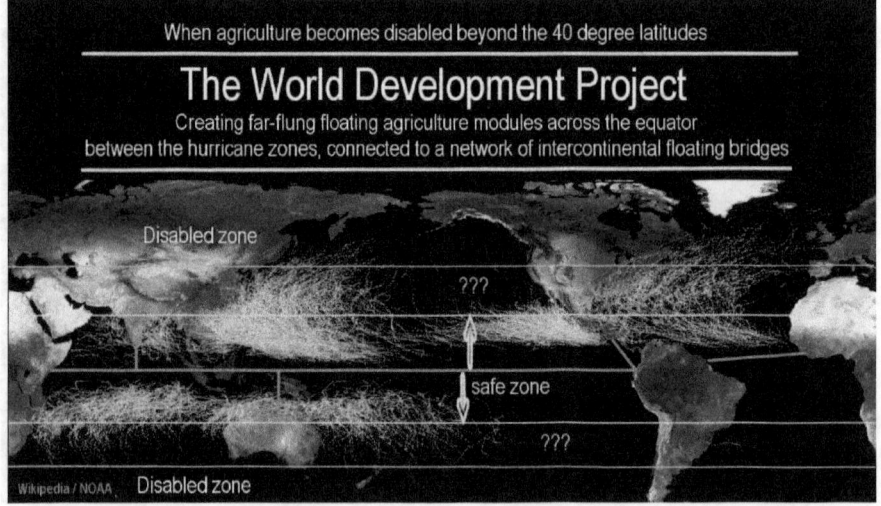

The cosmic-energy development, and Ice Age infrastructure development, are both elements of a single, global Science Challenge that no nation on the planet is exempted from. On this platform the challenge will be met as one of the most urgent common needs of all mankind, by which we stand united as one people of a single, universal, humanity.

www.ingramcontent.com/pod-product-compliance
Lightning Source LLC
Chambersburg PA
CBHW060358190526
45169CB00002B/658